THE

REMOUNT SERVICE IN THE UNITED KINGDOM

IN

WAR-TIME.

FireStep Publishing
Gemini House
136-140 Old Shoreham Road
Brighton
BN3 7BD

www.firesteppublishing.com

First published by the General Staff, War Office 1913.
First published in this format by FireStep Editions,
an imprint of FireStep Publishing, in association with
the National Army Museum, 2013.

www.nam.ac.uk

ISBN 978-1-908487-70-4

Cover design FireStep Publishing
Typeset by FireStep Publishing
Printed and bound in Great Britain

Please note: *In producing in facsimile from original historical documents, any imperfections may be reproduced and the quality may be lower than modern typesetting or cartographic standards.*

THE

REMOUNT SERVICE IN THE UNITED KINGDOM

IN

WAR-TIME.

(BEING A STATEMENT OF THE WAR-TIME ARRANGEMENTS AND THE DIRECTOR OF REMOUNTS' INSTRUCTIONS TO THE SERVICE UNDER HIS ORDERS.)

1913

www.firesteppublishing.com

THE

REMOUNT SERVICE IN THE UNITED KINGDOM

IN

WAR-TIME.

PART I.—THE REMOUNT SERVICE GENERALLY.

PART II.—THE REMOUNT DEPÔTS.

(B 2132) Wt. w. 3330—746 200 11/13 H & S P 13/387.

1. GENERAL ORGANIZATION.

1. The Remount Branch of the War Office in war-time has three functions. (For small wars special instructions will be issued.)

(*a*) The supply of horses to maintain the Base Remount Depôt of the Overseas Army.

(*b*) The maintenance after mobilization of the Territorial Force and other units that are mobilized in the United Kingdom.

(*c*) The control of the Remount Service in war in England.

2. The organization for the supply of horses in war-time is as follows :—

(*a*) On mobilization, as described in Chapter V of the Remount Regulations, the Reserve Units are completed to their War Establishment (*see* Part V, War Establishments) by impressment by the Command concerned, immediately the Expeditionary Force and Territorial Force are completed.

These consist of—

3 Reserve Brigades of Horse Artillery.
12 " " " Field Artillery.
14 " Cavalry Regiments.

These units receive the horses left behind by the mobilized units, and their establishments are completed by impressment. They train horses for drafts overseas. On drafts being required the Director of Remounts at the War Office orders them to be prepared and despatched by the Reserve Units to the port of embarkation.

(*b*) The four Remount Depôts that exist in the United Kingdom in peace-time, viz.—

No. 1. Dublin and Lusk,
" 2. Woolwich,
" 3. Melton Mowbray,
" 4. Arborfield Cross,

are expanded in war-time to an establishment of 500 horses each. As there are no reserve units for Army Service Corps, No. 4 Depôt acts as a reserve unit for Army Service Corps, and has an equipment of G.S. wagons for training young draught horses. These depôts are completed, so far as possible, on mobilization by impressment (as shown in Part V of the Remount Statement). They are refilled by remounts purchased in the United Kingdom (especially Ireland) and abroad, as the case may be.

(*c*) The mobilized Territorial and other Home Defence Forces receive their remounts in the form of drafts despatched by orders of the Director of Remounts at the War Office, from the depôts described above, or from Reserve Units. Each depôt will be equipped with a special detachable *personnel* for the establishment, if necessary, of temporary Remount Depôt for the Territorial Force.

(*d*) The reserve units contain six months' supply of cavalry horses, and three months' supply of other horses for the Expeditionary Force.

The Director of Remounts at the War Office controls the Remount Service of the Mobilized Home Defence Forces, as if the War Office was the headquarters of the I.G.C., and issues of horses from depôts are made on his order only, save in the event of operations in the United Kingdom.

3. The officers of the Remount Service in peace-time will receive special instruction as to their duties when mobilization is completed. One D.A.D.R. will be detailed for administrative duty at the headquarters of each command.

2. THE SUPPLY OF HORSES.

1. The mobilization of the Expeditionary and Territorial Forces and Reserve Units will, it is anticipated, leave Great Britain practically cleared of horses fit for military purposes that will be offered for sale. It will, however, complete the Reserve Units, as explained in the foregoing section, and it will also leave 500 horses in each of the four Remount Depôts.

2. In Ireland some 17,000 horses will have been taken for the Army. It is anticipated, however, that there will still be some thousands of horses to be bought there. The three Inspectors of Remounts, with assistants,* therefore, proceed to Ireland as soon as mobilization is complete, to purchase locally such horses as may be required to replace the drafts on the Reserve Units and the four Remount Depôts. The Inspectors will be relieved of all control of the depôts.

3. It will also probably be necessary to despatch remount commissions abroad to purchase horses in any markets that may be available. Two Remount Commissions of 4 officers each are held in readiness.*

PART II.—**THE REMOUNT DEPÔTS.**

SECTION 3.—GENERAL DESCRIPTION OF FUNCTIONS.

1. The four Remount Depôts in the United Kingdom, viz.—

No. 1. Lusk and Dublin,
 „ 2. Woolwich,
 „ 3. Melton Mowbray,
 „ 4. Arborfield Cross,

will be enlarged on mobilization and equipped with the establishment shown in Appendix I. This is taken from the War Establishments, Part —†. The depôts will be removed from the control of the Inspectors of Remounts, and will be commanded by

* *See* Appendix III.
† Not yet published.

a Deputy-Assistant Director of Remounts, under whom the peace-time Superintendent will work. (Sections 7 and 8.)

2. They will be filled by impressment by Commands, as detailed in the Remount Statement, to an establishment of 500 horses each, and will be primarily considered as accommodation depôts, in which horses receive the full peace ration.

3. Their duties are to fill the Reserve Units, and to issue remounts to Territorial and other units within the United Kingdom, in accordance with orders from the War Office. (*See* Section 1, para. 2 (*c*) and Section 3.)

4. The depôts will be filled, after they have been completed by impressment—

(*a*) By horses purchased in the United Kingdom, so far as there may be any available.

(*b*) By horses purchased in such countries oversea as may be open to us.

5. The depôt war-time equipment is detailed in Appendix II.

4. *PERSONNEL* ON MOBILIZATION.

1. The establishment of each depôt is detailed in Appendix I, and will consist of the military nucleus for control of office stores, and the general conducting of parties of horses, and the civilian *personnel* for the actual handling of the animals. A further duty will be to furnish supervising *personnel* for any temporary remount depôts opened to supply the Territorial Force during operation. (*See* Section 1, para. 2 (*c*).)

2. The majority of the military peace establishment join remount units of the field army on mobilization. The following nucleus of remount military *personnel* remain on mobilization, viz.—

1 Staff Serjeant,
2 Serjeants,
1 Q.M.S. or S.S. Farrier.

If the requirements of mobilization do not take the men equally from depôts, the remaining establishment will be adjusted between depôts under instructions from the War Office. Superintendents of depôts will report to the War Office the numbers of the military *personnel* remaining on mobilization. The remaining soldiers to complete war-time establishments will be detailed as may be arranged by the War Office.

3. The civilian *personnel* will be engaged by Superintendents and be paid at the usual peace rates.

5. ACCOMMODATION.

1. In the event of mobilization during the summer months, viz., between April and October, the horses in the depôts will run in the paddocks, except so far as stabling is available. Those for immediate issue to Reserve Units being brought up into stables.

2. Plans of temporary stables to be erected for winter accommodation exist at each depôt, and arrangements for erecting them exist at all Command Headquarters concerned.

3. The depôt equipment Tables include tentage for extra *personnel* and picketing gear for 100 horses, which latter will probably not be required. Such further issue of rugs as may be necessary will be made in winter-time.

6. STEPS TO BE TAKEN ON AND PRIOR TO MOBILIZATION. (*By the Superintendent.*)

Prior to Mobilization.

1. Prior to mobilization the following steps are necessary :—

(*a*) Submit indents for war-time equipment.

(*b*) Arrange places for the reception of the war-time establishment.

(*c*) Discuss with Command Headquarters the arrangements for erecting temporary stables referred to in Section 5 (2), if ordered.

(*d*) So arrange duties that the permanent nucleus referred to in Section 4 (2), can carry on the essential duties.

(*e*) Consider the steps necessary to engage the extra civilian establishment required, with their cooks.

(*f*) Select site for camp of extra *personnel* and their kitchens.

(*g*) Have proposals ready for commandeering extra adjacent land.

On Mobilization.

2. On mobilization the following steps will be taken :—

(*a*) Draw war-time equipment.

(*b*) Carry out the various steps detailed above.

(*c*) Inform Director of Remounts if there are any horses in depôts not likely to be fit for issue within six months.

7. DUTIES OF DEPÔT COMMANDANTS ON MOBILIZATION.

1. The Depôt Commandant is responsible for the general control of his depôt, which includes the following duties :—

(*a*) Selection of all horses for issue.

(*b*) Giving instructions to the Superintendent as to depôt routine.

(*c*) Making all reports to Director of Remounts.

(*d*) Preparation of any temporary field depôt required to be detached.

(*e*) Care (other than veterinary) of all horses in the depôt.

8. DUTIES OF THE SUPERINTENDENTS ON MOBILIZATION.

1. The Superintendent is in executive charge of the routine of the depôt, and his duties include the following :—

(*a*) Employment and control of civilian employés.

(*b*) Carrying out all depôt routine.

(*c*) Bringing to notice all matters requiring the Commandant's orders, or reference to the Director of Remounts.

(*d*) Command of military *personnel*, under the general orders of the Commandant.

APPENDIX I.

Establishment of a Remount Depôt (U.K.) in War-time for 500 Horses.

	Depôt.	Attached—detachable with horses.	Total.
Commandant	1	1	
Superintendent	1	—	
S.S. Serjeant-Major	1	—	
S.S.Q.M.S.	1	—	
Serjeant and Corporals ...	5	4†	
Q.M.S. or S.S. Farrier ...	1	2†	
Civilians.			
Nagsmen	25	—	
Foreman Studsman	5*	2‡	
Grooms	100*	40‡	
S.-Smiths	8	4‡	
Dressers	3	2‡	
Saddlers...	2	—	

* 1 foreman and 20 grooms to each 100 horses.

† Will be posted should there appear a likelihood of detached depôts for Home field troops being required.

‡ Will be entertained as in †

APPENDIX II.

EQUIPMENT REQUIRED BY A REMOUNT DEPÔT IN THE UNITED KINGDOM IN WAR-TIME.

Mauls, G.S.	20
Nets, hay	100
Pads, surcingle	100
Pegs, picketing, with loops	220
Ropes, head, hemp, with ring	400
Ropes, heel	200
Rugs, horse, large	100
,, ,, small	100
Surcingles, web, cavalry	100
Troughs, proof, 600 galls.	20
Shovels, G.S.	50
Posts, picket, 5 ft.	25
,, ,, 2½ ft.	50
Cordage, hemp, hawser, 3 strand 2½ fm.	200
Bags, nose. G.S.	150
Brushes, horse	100
Combs, curry	100
Rubbers, horse	250
Sponge, 7 dm.	100
Chests, tool, filled, No. 7	5
Forges, field, G.S., complete	3
Covers, sail cloth—33 ins. × 30 ins.	25
,, ,, 24 ins. × 18 ins.	25
Barrow, stable	10
Irons, branding, 5/8 } Figures, 0—8, sets }	10
Pails, I.G.	100
Tubs, coal (for feeding horses in paddock)	100
Tents, C.S.L., if required—1 per 6 civilians } ,, ,, 1 per 3 N.C.Os. }	37
Tents, marquee H.P. small (for messing, &c.)	10
G.S. wagons with double set of harness (*see* Sec. 1, para. 2 (b))	2*

APPENDIX III.

Detail of Officers held ready for remount duty on mobilization (*see* paper A.B. 1590). (Detailed by M.S., War Office.)

(A) *War Office Extra Staff.*
  2 Officers.
  1 Quarter-Master.
(B) *War Office Extra Purchasers* (to assist I. of Rs.).
  12 Officers.
(C) *War Office. Foreign Commission.*
  2 Commissions and 4 Officers each.

* No. 4 Depôt only.